YOUR KNOWLEDGE HAS VALUE

- We will publish your bachelor's and
  master's thesis, essays and papers

- Your own eBook and book -
  sold worldwide in all relevant shops

- Earn money with each sale

Upload your text at www.GRIN.com
and publish for free

Bibliographic information published by the German National Library:

The German National Library lists this publication in the National Bibliography;
detailed bibliographic data are available on the Internet at http://dnb.dnb.de .

Imprint:

Copyright © 2015 GRIN Verlag
Print and binding: Books on Demand GmbH, Norderstedt Germany
ISBN: 9783668754034

This book at GRIN:

https://www.grin.com/document/433465

Rainer Stickdorn

Water Analysis in field and lab (chromatography, AAS, IC, photometry)

GRIN Verlag

Course 3214: SM9 Water Analysis - exercises

Report

Rainer Stickdorn, Matr. 20 45 283

**2nd semester MSc TropHEE
Institut für Angewandte Geowissenschaften
SS 2015**

Table of Contents

1 Introduction

The Water Analysis exercises (TuCaN 3214) are part of Special Modul SM9 "Hydrogeological Methods" of the MSc TropHEE and scheduled for the 1nd semester but had to be adjourned to the 2nd semester due to capacity bottlenecks in the lab. This course prepares for the Hydrogeological Field Course (TuCaN 3417) scheduled for the 2st semester. The Water Analysis course contains lectures and a practical part with surface water sampling, measuring water temperature, EC, pH, oxygen concentration and alkalinity in the field as well ion concentrations in the lab with AAS, IC and Photometry. A salt concentration experiment (EC = f(salt concentration)) with 2 different salts and an introduction to chromatography was also part of the exercises.

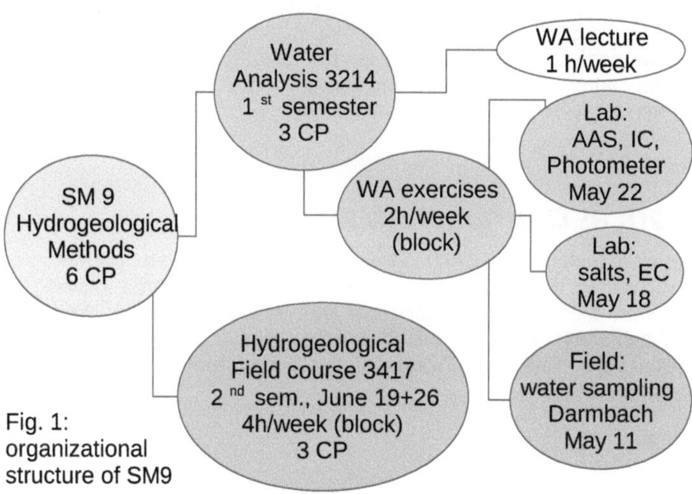

Fig. 1: organizational structure of SM9

TropHEE Module Handbook names as goals of this course: "The students are enabled to apply basic field techniques to characterize groundwater levels, groundwater flow fields, and to characterize aquifers in term of hydraulic properties. Students acquire methodical skills to use standard laboratory equipment to analyze water samples and to evaluate the results. Through the hands-on field and laboratory work they gain soft skills such as organizational skills, team working skills, communication skills, and data presentation skills."

The 25 TropHEE students were assigned to one of several groups that worked at different times on the different topics.

For our group 1 the work was divided into the following 3 days:

Monday, May 11 we walked along the stream Darmbach with a water sampling case with multimeter, several electrical probes, plastic bottles, titrator, ... At six stations we took water samples: 2 plastic bottles per station/location - one for anion-analysis and one for cation-analysis (with 1 cubic cm of acid HCl to stabilize the sample against degradation/precipitation before being analyzed in the lab) and measured water parameters like water temperature, EC, pH and oxygen concentration. Alkalinity (HCO_3^+, CO_3^{2+}) was measured by titration with 1.6 normal sulfuric acid (H_2SO_4) until the related indicator changes color (at pH 4.3)

Monday, May 18 we got an introduction to chromatography by Claudia Cosma and conducted an experiment dissolving increasing amounts of 2 salts and measuring the electric conductivity EC of the two solutions supervised by Dr. Marandi.

Friday, May 22 we analyzed our water samples from Darmbach with lab equipment like AAS, IC and Photometry supervised by Zahra Neumann. The samples from May 11 were stored in a fridge until opening the bottles in the lab on May 22.

2 Field work at Darmbach (May 11)

Water samples were taken alongside the stream Darmbach and water parameters like temperature, EC, pH, oxygen concentration, and alkalinity were measured in field. The samples from 6 stations were stored cool and dark until later analysis in the lab. We walked from the spring of Darmbach south east of Darmstadt (spring = station 1 = eastern end) into the city of Darmstadt west of Lake Woog (post Woog = station 6 = western end):

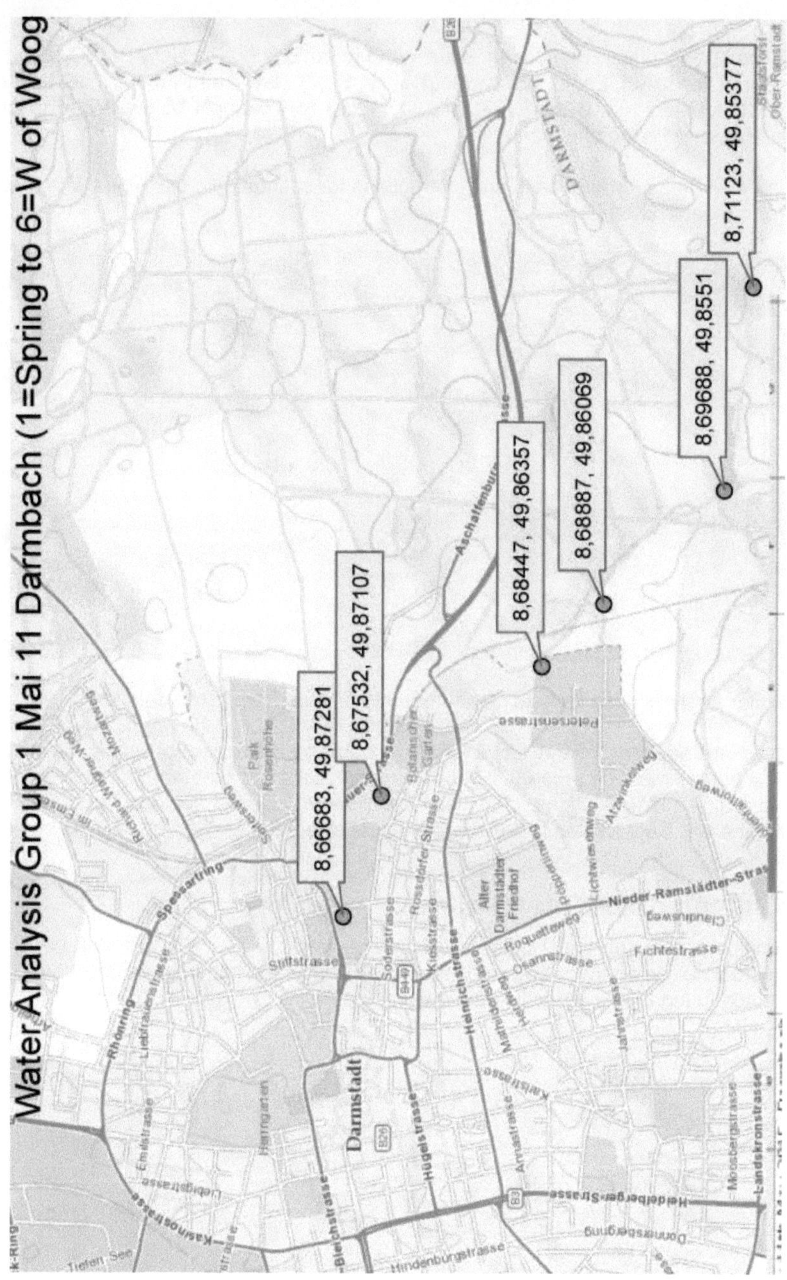

Fig. 3: Six sampling locations (western end = 6 = Post-Woog .. eastern end = 1 = spring) alongside stream Darmbach (produced w. ESRI ArcGIS).

We took two 100 ml plastic bottles per event – one for anions (F, Cl , NO_2, Br, NO_3, PO_4, SO_4) and one for cations (Mn, Fe, Li, Na, K, Mg, Ca, Sr, Si,...) stabilized with 1-2 ml of acid e.g. 2-normal HCl (= 2-molar = 2 mol HCl/L = 73g HCl/L) to lower pH below 2 and to prevent degradation (precipitation of metals).

We got a WTM measurement case and instructions for calibrating multimeters with probes by the lab for inorganic chemistry:

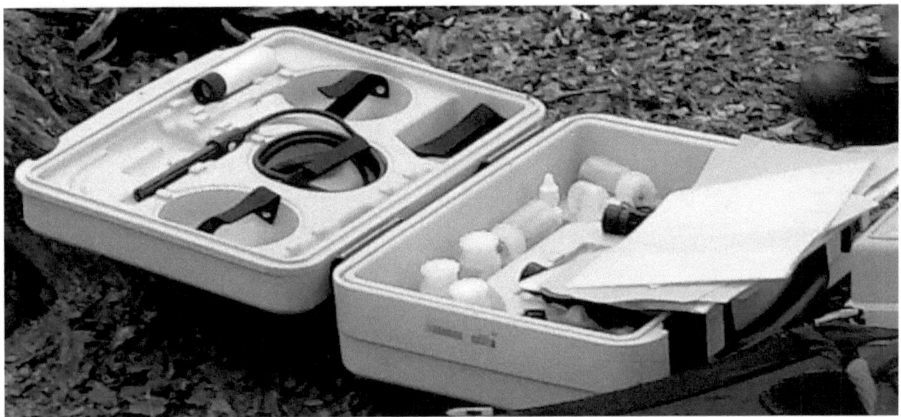

Fig. 4: Water analysis case for field measurements; picture from [Schiedek 2014].

The bottles are flushed several times with the water to be analyzed before the sample is taken. The bottles must be properly closed, marked/labeled so that they are clearly identified in the lab (ID, A_nion, C_ations, date, location, person, ...) and kept cool and dark before opened in the lab. Data related to the samples must also be documented in a field book.

2.1 Multimeter measurements

Fast changing parameters must be measured immediately in the field: temperature in °C, pH (negative decimal logarithm of H+ concentration), EC (electric conductivity in µS/cm), oxygen concentration in mg/L (and alkalinity due to escaping CO_2).

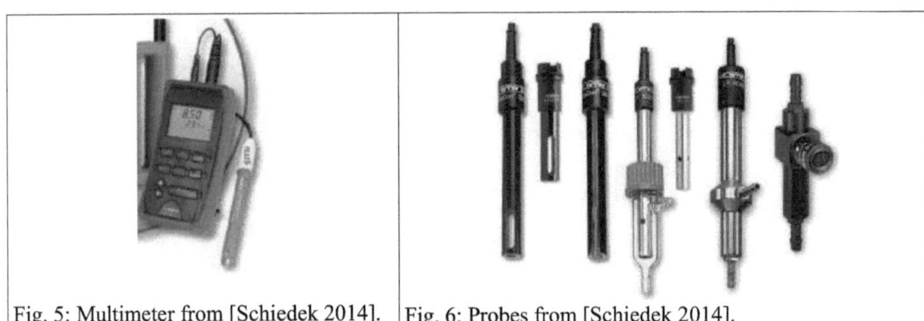

Fig. 5: Multimeter from [Schiedek 2014].   Fig. 6: Probes from [Schiedek 2014].

Ideally several probes sticking in a Flow-Through-Cell (where pumped water flows through) measuring most of the field parameters at the same time and place. Different probes can be connected to the same multimeter. Unfortunately we had no pump and no Flow-Through-Cell. So we had to do the measurements one after the other.

The multimeter converts automatically pH-measurements into pH-values at 25°C by computation. The pH-probe has a very thin glass electrode that must always be kept wet. It is stored in a 3-molar potassium chloride (KCl) buffer solution of pH=7 and must be calibrated for two pH-values with the expected measurements in the middle. Some devices give oxygen concentrations not only in mg/L but also relate the measurement to the saturation value at the current temperature and give the oxygen (saturation index) also as %. We got the following results:

2.2 Titration for alkalinity

Alkalinity is the capacity of an aqueous solution (of our water sample) to neutralize an acid. Alkalinity depends mainly on carbonates, i.e. on the concentration of bicarbonate (HCO_3^-) and CO_3^{2-} ions. Their concentration depends on the carbon dioxide (CO_2) concentration. Since CO_2 evaporates fast, also the alkalinity must be measured fast. This is done by titration with 1.6 normal sulfuric acid (1.6 equ./L = 0.8 mol/L of H_2SO_4): 100 ml of fresh water are taken into a glass bulb/flask. A Bromide-Cresol-Green-Methyl-pH-Indicator is added to detect the color change at pH 4.3. The sampled water is a little bit basic with a pH>7. Adding acid lowers the pH. The amount of acid added until color changes from green to pink is proportional to alkalinity. Always mixing all ingredients by moving the flask around we added slowly the acid by turning the wheel of a micro titrator (Hach Digital-Titrator) releasing small drops of acid.

100 ml water sample

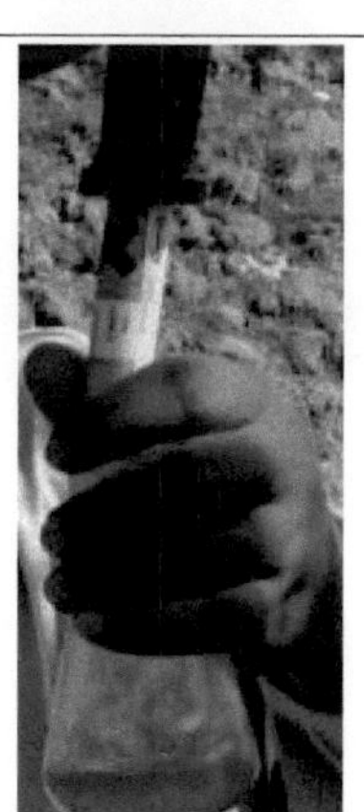

Fig: 7: Titration (from script) starts with green color.

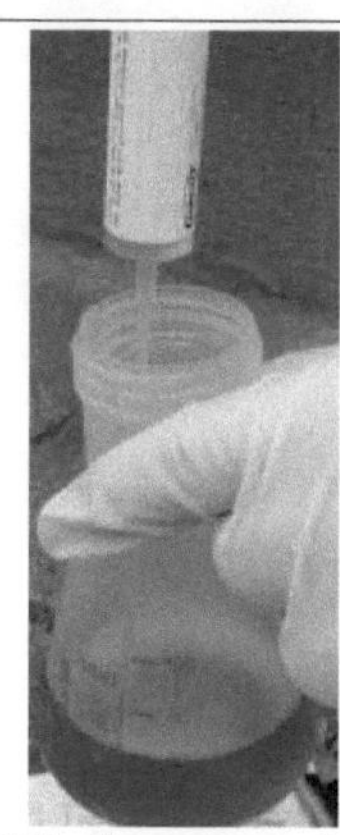

End of titration when color changes from green to pink

The titrator counts the drops of acid. In our case (100 ml water, 1.6-normal acid, certain drop size) the number of drops is equal to the alkalinity in mg $CaCO_3$ per liter. We had to assemble titrator, acid cartridge and delivery tube, to remove air and to set the counter to zero before first use:

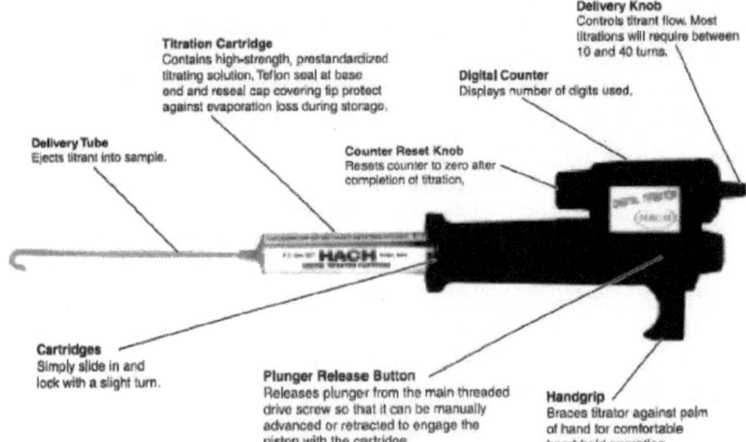

Fig. 8: Titrator with acid cartridge and delivery tube; picture from manufacturer.

2.3 Results from field measurements

Multimeter measurements and titration of the surface water at six locations at Darmbach delivered the following results:

	A	B	C	D	E	F	G
1	Group 1 on monday May 11th 2015, Darmbach, water sample size = 100 ml						
2							
3	Number	6	5	4	3	2	1
4							
5	Lattitude °N	49.87281	49.87107	49.86357	49.86069	49.8551	49.85377
6	Longitude °E	8.66683	8.67532	8.68447	8.68887	8.69688	8.71123
7	Location	Post-Woog (W of Woog)	Pre-Woog (E of Woog)	Vivarium	railway station Lichtwiese	Post Fish Pont	Darmbach spring, 2inch pipe
8	Hour:Min	12:30:00	11:53:00	15:07:00	13:07:00	13:40:00	14:10:00
9							
10	Water-Temperature in °C	19	14.1	16	15.6	18.7	9.4
11	pH at 25°C	8.08	7.9	7.9	7.94	7.72	7.26
12	EC in micro-S/cm	383	449	410	405	340	482
13	Oxygen in mg/l	7.87	9.37	9.4	9.15	7.88	8.3
14	titration w 1,6N H2SO4 in drops	195	192	185	180	142	210
15	Alcalinity (as mg/L CaCO3),	195	192	185	180	142	210
16	multiplier =1 for 100ml water						

Fig. 9: Field measurements from six stations at Darmbach on May 11 (6 = Post-Woog, 1 = spring).

3 Chromatography & salt experiment (May 18)

3.1 Chromatography

With Chromatography chemical mixtures are separated by injection system, chromatographic column and a solvent (liquid or gas) that transports the components of the mixture with different speeds through the column, so that they leave the column at different times: different substance is converted into different time of leaving the column.

The transport speed of every compound depends on physico-chemical properties like (de-)sorption to the filling material (stationary phase) in the column and the solubility of the compound into the solvent (mobile phase, GC=Gas Chromatograph, HPLC = High Performance / Pressure Liquid phase Chromatograph). High adsorption slows a substance. High solubility makes a substance faster. The retention time = time a substance is retained by adsorption – desorption processes is an identifying property of a substance.

The original sample (e.g. soil) is often not adequate for injection into a chromatograph. First the interesting substances must be extracted, filtered, enhanced in concentration, … An Accelerated Solvent Extractor (Soxlet) – ASE – that treats dry soil with hot solvents (acetone) was shown to us. After distillation small amounts of samples appropriate for injection (e.g. with a 1 µL syringe) into the chromatograph are provided. Other extraction methods are e.g. Solid-Phase Extraction – SPE – where the sample is given to a solid-phase agent that only binds one component that we want to to analyze and let all other components being washed away. An other solvent = eluant extracts than the component we want analyze from the solid phase.

The concentrations of components leaving the chromatographic column at different times must be measured by a detector. If further separation (more sensitivity) is needed a mass spectrometer – MS – can be placed between chromatograph and detector. MS uses a magnetic field that deflects flying charged particles (compounds are converted into charged ions by heat) according to their charge/mass ratio m/q. Modern MS use a quadropole mass analyzer – QMS – consisting of two pairs (=4) parallel rods energized by adjustable high frequency (RF = radio frequency) alternating currents. The particles try to travel along the rods but only such of certain m/q make it to the detector others are kicked out of the way. The ions reaching the detector arrive there at different times having different times of flight between the quadrupole rods. We have seen an 5975 from Agilent Systems as GC/MS-System. The detector can be a charge multiplier. Other detectors are:

TCD
The Thermal Conductivity Detector detects changes in thermal conductivity of the flowing gas of an GC due to components separated in the column before. Conductivity peaks show type of component (from time when peak appear) and concentration (from height of peak or area below the peak). This detector destroy nothing but is also not very sensitive.

FID
The Flame Ionization Detector burns components (destroys them) in an air-hydrogen flame to produce ions that flow to electrodes of opposite charge and build so an electric current that is measured. FID is only applicable for burnable substances like hydrocarbons and destroys them. So it should be the very last detector in a chain of different detectors.

ECD
The Electron Capture Detector measures the electric current produced by e.g. Ni-63 as radio-active β-emitter (electron emitter) between 2 electrodes and attenuated by electron capturing

components (flowing between the 2 electrodes) those concentrations are detected by the decrease of the electric current. ECD is very sensitive compared to TCD and TID but a less wide dynamic range.

The following picture shows the covered concentration range of some detectors:

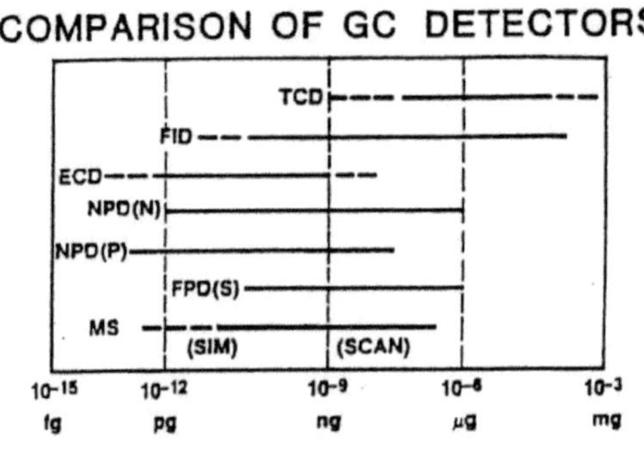

Fig. 10: Concentration ranges of detectors; from [Schiedek 2014].

A GC from Agilent Technologies was shown to us. Chromatographs often have an auto-sampler = carousel with many sample flasks with rubber septum so that a "robot arm" with syringe can take a µL from them (from the liquid or from the gas in the space in the flask above the liquid) and inject it through a septum of the chromatograph. Liquid taken from sample is evaporated on hot glass wool at 230°C and travels with helium as gas (= mobile phase) through the long column. The glass column of only 0.25 mm diameter (looks like a thin blank copper wire) is 10-100 m long and winds up to a coil of 20 cm in diameter. This coil is located in an electric oven that keeps temperatures between 40 and 300 °C. At the end of the column different detectors can be attached. Electrical signals from detectors are registered and processed by a computer. Concentration peaks over time for the investigated samples and for known standards are plotted / displayed by the equipment. For example the peak for PCE will be later than that for TCE that is later than that for Methanol.

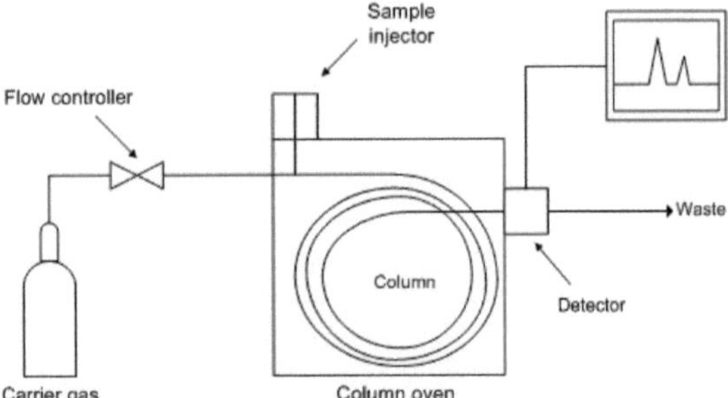

Fig. 11: Gas Chromatograph; from english wikipedia, upload by Offnfopt.

3.2 Salt experiment

We dissolved NaCl (halit, table salt, kitchen salt) and NaHCO3 (sodium bicarbonate, sodium hydrogen carbonate, baking soda) in water and increased stepwise the concentration of the two solutions from 20 to 5000 mg in 1 liter water and measured the electric conductivities EC of the two solutions for every step. The Multimeter with EC-probe measured EC in µS/mol.

What EC-values we have to expect is computable. Ions have an equivalent ionic conductance EIC in cm² S/mol. Molar EIC of some ions were given at a temperature of 25°C (lab temperature) in the instructions for this experiment. In the following table the EIC per gram and EIC of molecules are calculated:

matter	Molar weight in mg/mmol	EIC in cm*cm*mS/mmol	EIC in cm*cm*µS/mg
Na+	23	50	
Cl-	35.5	76.3	
NaCl	58.5	126.3	1000*126.3/58.5 = 1000***2.159**
HCO_3-	61	44.5	
$NaHCO_3$	84	94.5	1000* 94.5/84 = 1000***1.125**
Ca++	40.1	59	
$CaHCO_3$	101.1	103.5	1000*103.5/101.1=1000*1.024
$Ca(HCO_3)_2$	162.1	148	1000*148/162.1=1000***0.913**

Fig. 12: Molar EIC (from experiment instructions) converted into per-mg-EIC by division with molar weights.

We can expect that kitchen salt has nearly twice the effect of soda (2.159 vs 1.125). With salt concentration c = salt mass 'm' per volume V=1L=1000 cm³ we calculate EC:

EC= c * EIC = c * EIC = m * EIC / 1L   [1]
EC/(µS/cm) = EIC/ 1000cm³ * m_salt/mg   for V=1L   [2]
EC/(µS/cm) = 2.159 * m_NaCl/mg   [3]
EC/(µS/cm) = 1.125 * m_NaHCO3/mg   [4]

Theoretically EC is direct proportional to the EIC in cm² µS/mg and to the salt mass 'm' in mg (in one L). The following table contains salt amounts, measured and with equation [3] and [4] calculated EC in µS/cm:

	A	B	C	D	E
1	Salt mg / L	EC NaCl microS/cm	EC NaHCO3	EC_calc NaCl	EC_calc NaHCO3
2	0	14.2	13	0	0
3	20	110.2	58.6	43.18	22.5
4	50	191	95.2	107.95	56.25
5	100	324	175	215.9	112.5
6	200	505	310	431.8	225
7	500	1070	710	1079.5	562.5
8	1000	1968	1320	2159	1125
9	2000	3800	2500	4318	2250
10	5000	8700	5660	10795	5625

Fig. 13: EC-measurements and -calculations for 2 salt-solutions.

I exported the spreadsheet above as text file with tab-separated columns '2salts.csv' and put a #-sign in front of the header line to convert it to a remark in GNUplot.

```
#Salt mg / L   EC NaCl microS/cm  EC NaHCO3   EC_calc NaCl EC_calc NaHCO3
0      14.2   13     0      0
20     110.2  58.6   43.18  22.5
50     191    95.2   107.95 56.25
100    324    175    215.9  112.5
200    505    310    431.8  225
500    1070   710    1079.5 562.5
1000   1968   1320   2159   1125
2000   3800   2500   4318   2250
5000   8700   5660   10795  5625
```

There are 5 tab-separated columns. I tell GNUplot with a using-clause which column to use for which of the 4 graphs (1+2, 1+3, 1+4, 1+5). The calculated values are plotted with lines and points (lp = linepoints) the measured only with lines (w l). NaCl is plotted in red (lc = line color) and NaHCO$_3$ in blue. The legend is printed at the right margin (rmargin):

```
# EC = f(salt concentrations) Lab for group1 on May 18

set title 'electric conductivity EC in microS/cm over salt concentration in mg/L'
set xlabel 'concentration in mg/L'
set ylabel 'EC in microS/cm'
set key rmargin
plot [0:5000] [0:11000] "2salts.csv" \
        using 1:4 w lp lc 'red'  title 'NaCl calculated',\
     "" using 1:2 w l  lc 'red'  title 'NaCl measured',\
     "" using 1:3 w l  lc 'blue' title 'NaHCO3 measured',\
     "" using 1:5 w lp lc 'blue' title 'NaHCO3 calculated'
```

The GNUplot commands above in program '2salts_a.plt' produce the following picture:

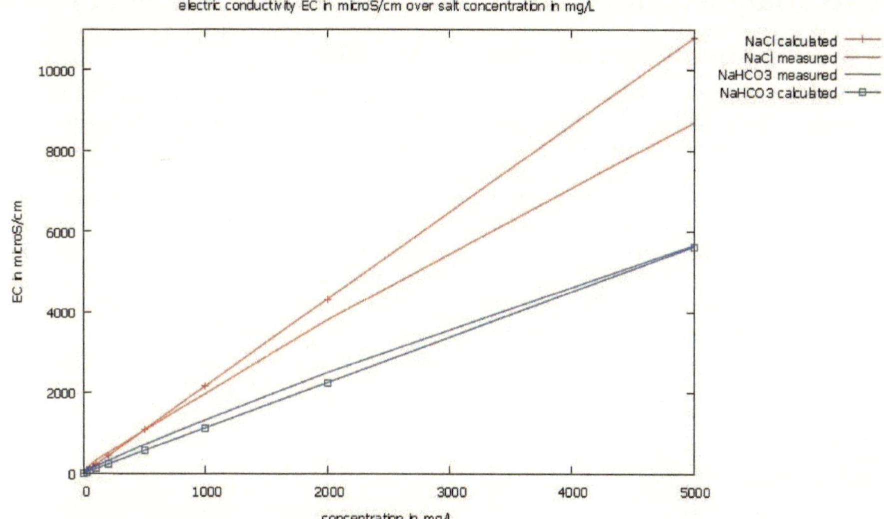

Fig. 14: EC = f (salt concentration) for 2 salts by GNUplot.

The questions from the experiment script are answered in the following way:

I. The expectation that NaCl increases much stronger (nearly twice due to EIC ratio of 2.159 / 1.125) the EC is fulfilled also by the measurements. We find the largest differences between measured and calculated EC for small and for large salt concentrations. Due to other ions in the water (also before adding any salt) the EC is greater than 0 for a salt concentration of 0. For high concentrations of NaCl the measured EC lacks behind the calculated EC because collisions between ions are more likely for higher concentrations and collisions slow down the flow of ions (the electric current) in the electric field of the multimeter probe what is equivalent to adding resistance. For NaHCO3 the EC is not high enough to show the same effect.

II. Water mainly with NaCl (from sea water intrusion) and EC=500 µS/cm has roughly a TDS of 200 mg/L (see line 6 column A and B in the spreadsheet)

III. Water from an inland carbonate aquifer (i.e. with Ca-HCO3 type - means $Ca(HCO_3)_2$ instead of $CaHCO_3$?) and EC=500 µS/cm has a TDS higher than that for $NaHCO_3$ due to the lower EIC of 1000* 0.931 cm* cm* µS/mg ==>

EC/(µS/cm) = 0.931 * m_Ca(HCO3)2 /mg   [5]
=> mass in mg per L = EC/0.931=500mg/0.931=537

IV. I have learned that different ions have different molar EIC and therefore also EIC per mg. For low concentrations the EC is higher than that calculated from the added salt because there are already unknown ions in the water and for high concentrations the EC is lower than that calculated from added salt, because there are more collisions in a stronger current that slows down charged particles (means more resistance = less EC).

4 Lab analysis with AAS, IC and Photometry (May 22)

4.1 Atomic absorption spectroscopy - AAS

AAS – detects elements (metals) that are also used in a special electric (anode + cathode) lamp, where the metal is burned in a hollow cathode to produce a specific optical spectrum. Alternatively special adjustable lamps are used that produce the spectrum needed to detect certain metals without burning the same metal in the cathode of the lamp. TU-DA uses the later approach with a Silon-UV-lamp. The water sample is vaporized and the solids in the sample are burned in a laterally long flame producing a cloud of metal atoms that can interact with certain optical wavelengths in the lamps light. A light beam from the lamp is directed through the flame and specifically attenuated by the metal atoms in the gas from the sample. Opposite to the lamp at the other end of the long flame an optical detector measures the remaining intensity of certain wavelengths.

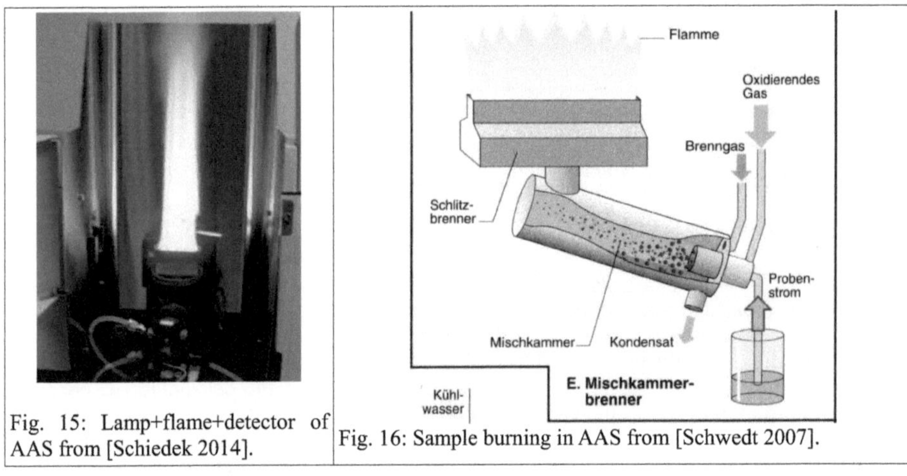

Fig. 15: Lamp+flame+detector of AAS from [Schiedek 2014].

Fig. 16: Sample burning in AAS from [Schwedt 2007].

In the contrAA-machine of TU-DA an echelle (diffraction) grating (many parallel grooves in distances similar to lights wavelength) works similar to an optical prism in splitting light (coming from xenon lamp and passing the sample) into all its different wavelengths.

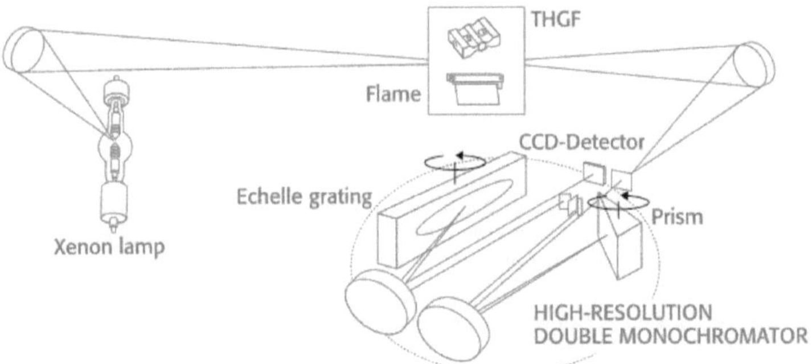

Fig. 17: Optical path in AAS, picture from manufacturer analytikJena (of contrAA-300).

The Darmbach water was analyzed for Fe- and Mn-ions with AAS. The Silon lamp provides a wavelength of 248 nm for Fe- and of 279 nm for Mn-Ion-measurements (for 4 seconds per sample). The following concentrations were measured:

	Fe	Mn
1=Spring	0.0152 mg/L	0.0039 mg/L
2=After Fish Pont	0.1419 mg/L	0.1521 mg/L
3=Railway station/bridge	0.3303 mg/L	0.2364 mg/L
4=Lichtwiese / Vivarium	0.2845 mg/L	0.2095 mg/L
5=Pre-Woog Lake	0.2506 mg/L	0.1758 mg/L
6=Post-Woog Lake	0.1394 mg/L	0.1193 mg/L

Fig. 18: Fe- and Mn- concentrations measured by AAS – values at 1=spring are detection limits, i.e. real values are below the detection limit.

4.2 Ion Chromatography - IC

Ion exchange chromatography uses a liquid as mobile phase as transport media for the sample's substances (analyte) that must be polar molecules or ions. These are retained on the stationary phase (resin or gel matrix of agarose or cellulose beads) in the column and finally eluted again by increasing concentrations of similarly charged species displacing the analyte. In cation chromatography cations (positive charged ions, e.g. metal ions) the analyte is eluted by (also positive charged) sodium ions. Different components of the analyte are eluted at different times and appear at the end of the column at different times.

At the end of the chromatography column a detector (e.g. TCD using thermal conductivity, FID using current from ions produced in a flame, ECD using electron capture by analyzed components decreasing electric current from radioactive β-emitter, MS using deflections dependence on charge and mass in magnetic fields, … using attenuation of light) measures the concentrations of the separated ions. Detectors were already described in the chapter about chromatography.

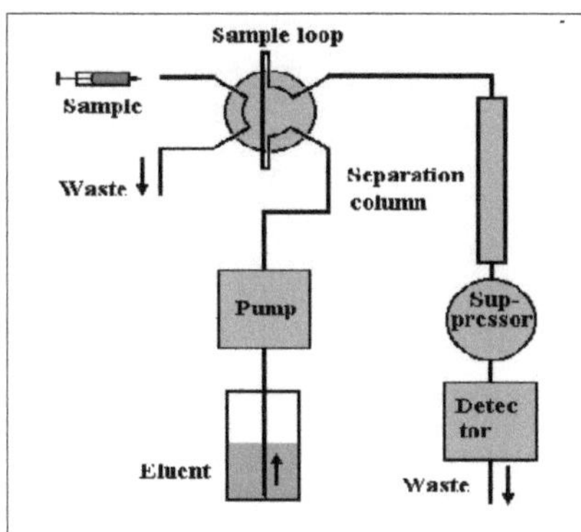

Fig. 18: IC principle from [Schiedek 2014].

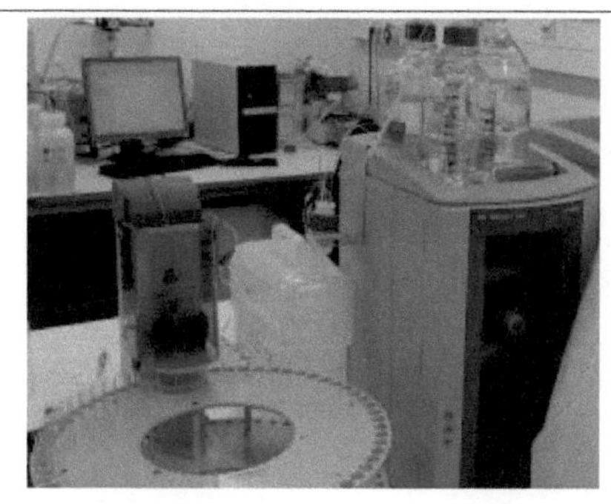

Fig. 19: IC with 882 Compact IC plus from Metrohm, photo taken by author.

With IC we measured the following ion concentrations in Darmbach water:

	A	B	C	D	E	F	G	H
	Ident/Location	F mg/l	Cl mg/l	NO2 mg/l	Br mg/l	NO3 mg/l	PO4 mg/l	SO4 mg/l
1	Ident/Location	F mg/l	Cl mg/l	NO2 mg/l	Br mg/l	NO3 mg/l	PO4 mg/l	SO4 mg/l
2								
3	Post Woog Lake	0.134	9.95		0.844	0.219		22.232
4	Pre Woog Lake	0.319	10.058	0.071	1.04		0.05	24.226
5	Lichtwiese/Vivarium	0.055	6.001	0.041	0.966		0.078	24.844
6	rail way bridge	0.047	4.395	0.041	0.84		0.067	19.167
7	After Fish pond	0.106	4.55		0.819	0.127	0.078	19.275
8	Spring	0.051	4.602	0.003	1.236	9.385	0.088	25.99

	A	I	J	K	L	M	N	O	P
1	Ident/Location	Li mg/l	Na mg/l	NH4 mg/l	k mg/l	Mg mg/l	Ca mg/l	Sr mg/l	dilution
2									
3	Post Woog Lake	0.007	10.177	0.177	2.29	5.58	58.984	0.175	2.5
4	Pre Woog Lake	0.01	9.979	0.159	2.535	6.328	72.412	0.223	2.5
5	Lichtwiese/Vivarium	0.01	8.783	0.086	2.012	6.193	65.641	0.15	2.5
6	rail way bridge	0.009	8.059	0.113	1.973	5.779	59.62		2.5
7	After Fish pond	0.008	8.407	0.143	2.336	4.749	53.937	0.135	2.5
8	Spring		5.966	0.143	1.448	7.398	80.559	0.215	2.5

Fig. 20: IC-results for (undiluted) Darmbach water.

The Darmbach water had to be diluted 1:2.5 to keep ion concentrations small enough to be measured precisely. The IC-machine is informed about this dilution so it gave the correct concentrations of the undiluted water of Darmbach.

4.3 Photometry

Photometry is an optical method where the absorption/attenuation of a light beam of known wavelength by the investigated material is measured. Attenuation of light is dependent on kind of penetrated material, its concentration and the lights wavelength. In Darmbach water the Si^{4+}-concentration was measured by the (double beam) photometer "Specord 200 plus" from analytikJena:

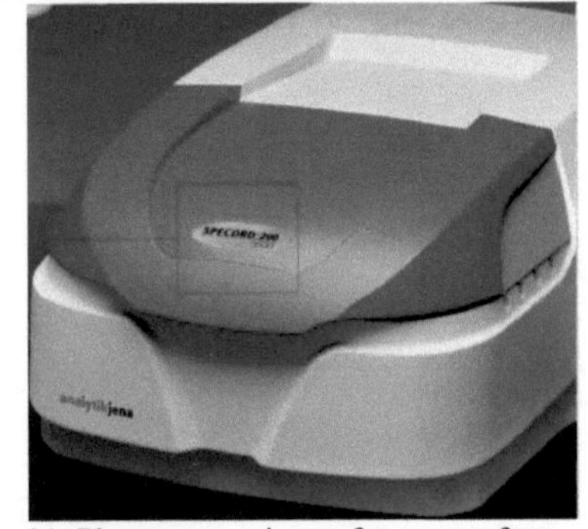

Fig. 21: Photometer; picture from manufacturer.

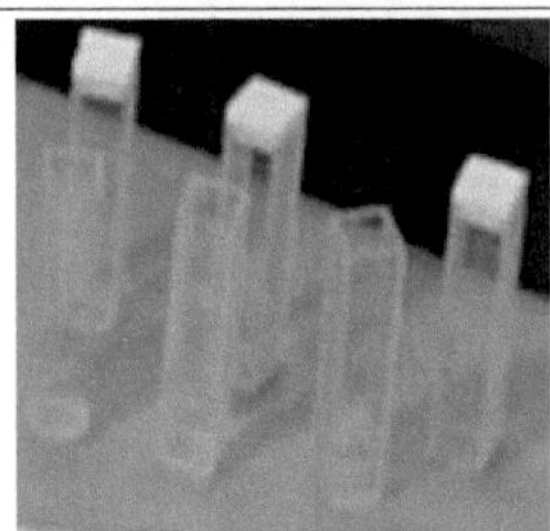

Fig. 22: Sample glasses to be inserted in Specord.

The Darmbach samples were accompanied by some standards of known Si-concentrations. The water is filled into small cubic flasks and one flask after the other is put into the machine to measure the attenuation. Afterwards the machine automatically adjusts the values computational according to the standards' values and used dilutions (read by the machine)

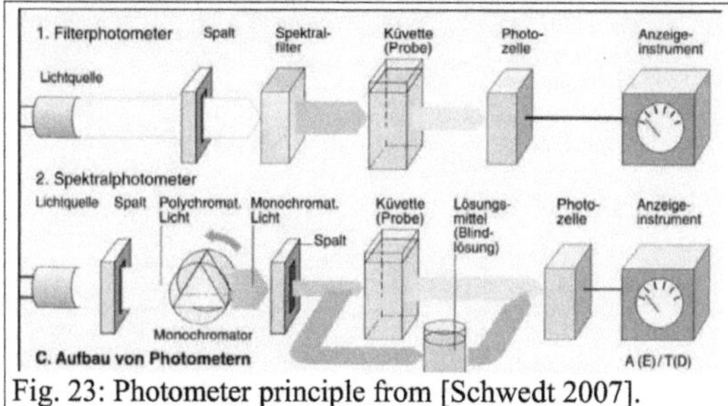

Fig. 23: Photometer principle from [Schwedt 2007].

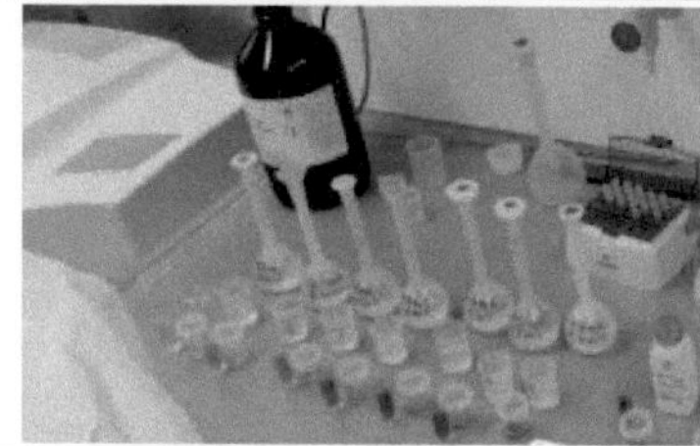
Fig. 24: Sample preparation for photometer; photo taken by author.

#	Location	Si mg/L	Remark
1	Spring	7.56	1:2 diluted
2	Fish Pond	3.15	1:2 diluted
3	Rail way station/bridge	6.83	1:2 diluted
4	Lichtwiese/Vivarium	6.83	1:2 diluted
5	Pre-Woog Lake	6.96	1:2 diluted
6	Post-Woog Lake	0.23	1:2 diluted
7	"	0.31	undiluted

Fig. 25: Si^{4+} -concentrations of (undiluted) Darmbach water.

5 Synopsis of measurements

With the ion concentration from titration (alkalinity) and from lab we can draw Schoeller- (all locations in one graph, x-axis = different ions, y-axis = concentrations), Piper- (2 triangle and 1 diamond diagram) and Stiff- (cations vs anions, 1 diagram per location) diagrams and start analyzing the data. Most diagram software is commercial (AqQA from RockWare, AquaChem from Schlumberger, ChemPoint from Starpoint, Geochemist's Workbench from Aqueous Solutions, EnviroInsite from HydroAnalysis). The question is: How does Darmbach water change from spring to the city of Darmstadt or vice versa (because most of the data is already ordered in this direction and the spring is east of the city i.e. in "reading direction")?

5.1 Schoeller diagram & ion balance

For a Schöller diagram we need the spreadsheet/matrix with IC-results transposed (columns converted into rows and vice versa). 2 rows for Fe- and Mn-concentrations from AAS, one for the Si-concentration from photometry and finally another one for the alkalinity from field measurements are added. All concentrations are still in mg/L. They are converted to meq/L by multiplying with valence/molar weight. Sum of meq/L for all cations vs all anions is called the ion balance and should give nearly equal numbers, but we get high errors (100%*(cations-anions)/(cations+anions)) of up to 10%:

	A	B	C	D	E	F	G	H
1	#	Conv2meq/L	Post Woog Lake	Pre Woog Lake	Lichtwiese/Vivarium	rail way bridge	After Fish pond	Spring
2	F	0.052631579	0.134	0.319	0.055	0.047	0.106	0.051
3	Cl	0.028169014	9.95	10.058	6.001	4.395	4.55	4.602
4	NO2	0.02173913	0	0.071	0.041	0.041	0	0.003
5	Br	0.0125	0.844	1.04	0.966	0.84	0.819	1.236
6	NO3	0.016129032	0.219	0	0	0	0.127	9.385
7	PO4	0.031578947	0	0.05	0.078	0.067	0.078	0.088
8	SO4	0.020833333	22.232	24.226	24.844	19.167	19.275	25.99
9	Li	0.142857143	0.007	0.01	0.01	0.009	0.008	0
10	Na	0.043478261	10.177	9.979	8.783	8.059	8.407	5.966
11	NH4	0.055555556	0.177	0.159	0.086	0.113	0.143	0.143
12	k	0.025641026	2.29	2.535	2.012	1.973	2.336	1.448
13	Mg	0.082304527	5.58	6.328	6.193	5.779	4.749	7.398
14	Ca	0.05	58.984	72.412	65.641	59.62	53.937	80.559
15	Sr	0.02283105	0.175	0.223	0.15	0	0.135	0.215
16	Fe	0.035714286	0.1394	0.2506	0.2845	0.3303	0.1419	0.0152
17	Mn	0.036363636	0.1193	0.1758	0.2095	0.2364	0.1521	0.0039
18	Si	0.142857143	0.23	6.96	6.83	6.83	3.15	7.56
19	alk	0.02	195	192	185	180	142	210
20								
21								
22	cation meq		3.9772081321	5.6798177593	5.2413117359	4.8726807824	3.9961391322	6.04241
23	anion meq		4.6540332465	4.646400698	4.3919834816	4.1277047906	3.3798220064	5.02793
24	Cat-an/(cat+an)		0.0784157324	0.1000770094	0.0881659116	0.0827715641	0.0835575343	0.09164
25	% Fehler		7.8415732418	10.0077009361	8.8165911571	8.2771564144	8.3557534284	9.16399

Fig. 26: All measurements (in mg/L) united with 1 column per location & ion balance.

Larger concentrations are divided by 10, 100 or 1000 to bring all values into the range 0... 1:

	A	B	C	D	E	F	G	H
1	#		Post Woog Lake	Pre Woog Lake	Lichtwiese/Vivarium	rail way bridge	After Fish pond	Spring
2	F	1	0.134	0.319	0.055	0.047	0.106	0.051
3	.1Cl	2	0.995	1.0058	0.6001	0.4395	0.455	0.4602
4	NO2	3	0	0.071	0.041	0.041	0	0.003
5	Br	4	0.844	1.04	0.966	0.84	0.819	1.236
6	.1NO3	5	0.0219	0	0	0	0.0127	0.9385
7	PO4	6	0	0.05	0.078	0.067	0.078	0.088
8	.01SO4	7	0.22232	0.24226	0.24844	0.19167	0.19275	0.2599
9	Li	8	0.007	0.01	0.01	0.009	0.008	0
10	.1Na	9	1.0177	0.9979	0.8783	0.8059	0.8407	0.5966
11	NH4	10	0.177	0.159	0.086	0.113	0.143	0.143
12	.1k	11	0.229	0.2535	0.2012	0.1973	0.2336	0.1448
13	.1Mg	12	0.558	0.6328	0.6193	0.5779	0.4749	0.7398
14	.01Ca	13	0.58984	0.72412	0.65641	0.5962	0.53937	0.80559
15	Sr	14	0.175	0.223	0.15	0	0.135	0.215
16	Fe	15	0.1394	0.2506	0.2845	0.3303	0.1419	0.0152
17	Mn	16	0.1193	0.1758	0.2095	0.2364	0.1521	0.0039
18	.1Si	17	0.023	0.696	0.683	0.683	0.315	0.756
19	.001alk	18	0.195	0.192	0.185	0.18	0.142	0.21

Fig. 27: All concentrations normalized to a range of about 0..1.

This table is also saved as tab-separated text file 'all_data10.csv' with a GNUplot remark sign (#) at the begin of the first line (= header line):

#		Post Woog Lake	Pre Woog Lake	Lichtwiese/Vivarium	rail way bridge	After Fish pond	Spring
F	1	0.134	0.319	0.055	0.047	0.106	0.051
.1Cl	2	0.995	1.0058	0.6001	0.4395	0.455	0.4602
NO2	3	0	0.071	0.041	0.041	0	0.003
Br	4	0.844	1.04	0.966	0.84	0.819	1.236
.1NO3	5	0.0219	0	0	0	0.0127	0.9385
PO4	6	0	0.05	0.078	0.067	0.078	0.088
.01SO4	7	0.22232	0.24226	0.24844	0.19167	0.19275	0.2599
Li	8	0.007	0.01	0.01	0.009	0.008	0
.1Na	9	1.0177	0.9979	0.8783	0.8059	0.8407	0.5966
NH4	10	0.177	0.159	0.086	0.113	0.143	0.143
.1k	11	0.229	0.2535	0.2012	0.1973	0.2336	0.1448
.1Mg	12	0.558	0.6328	0.6193	0.5779	0.4749	0.7398
.01Ca	13	0.58984	0.72412	0.65641	0.5962	0.53937	0.80559
Sr	14	0.175	0.223	0.15	0	0.135	0.215
Fe	15	0.1394	0.2506	0.2845	0.3303	0.1419	0.0152
Mn	16	0.1193	0.1758	0.2095	0.2364	0.1521	0.0039
.1Si	17	0.023	0.696	0.683	0.683	0.315	0.756
.001alk	18	0.195	0.192	0.185	0.18	0.142	0.21

This text file above is the input to GNUplot program 'schoeller_a.plt':

```
# Schoeller diagramm x= F,Cl,NO2, ...,alk; y= concentrations in mg/L
# per location (at Darmbach) one curve

reset
set title 'Schoeller diagram: concentrations in mg/L in water from Darmbach on May 11, 2015'
set xlabel 'ions: F, Cl, NO2, ..., Si, alk(alinity)'
set ylabel 'concentrations in mg/L, .1Cl=0.1*[Cl],\
 .01SO4=0.01*[SO4],..., .001alk=0.001*[alkalinity]'

set key rmargin
# x = column 2 but xticlabel(1) from column 1 is displayed at position x
plot "all_data10.csv" \
        using 2:3:xtic(1) w lp  lc 'red'  title 'Post Woog Lake',\
     "" using 2:4:xtic(1) w l   lc 'blue'  title 'Pre Woog Lake',\
     "" using 2:5:xtic(1) w lp  lc 'green' title 'Lichtwiese/Vivarium',\
     "" using 2:6:xtic(1) w l   lc 'yellow' title 'Rail way bridge/station',\
     "" using 2:7:xtic(1) w lp  lc 'black' title 'After Fish pond',\
     "" using 2:8:xtic(1) w l   lc 'pink' title 'spring'
```

The GNUplot program 'schoeller_a.plt' plots the following Schoeller diagram:

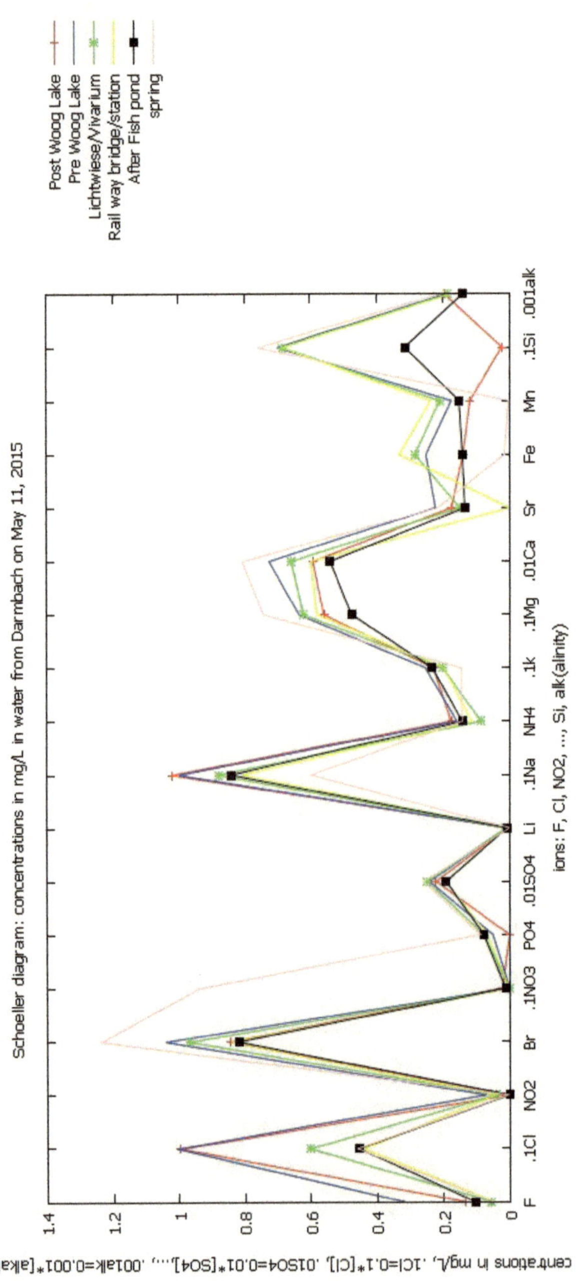

Fig. 28: Schoeller diagram for Darmbach water from May 11, 2015.

5.2 Stiff diagrams

We could plot Stiff diagrams with GNUplot reading x-coordinates as equivalent (meq) concentrations (concentration * valence) with negative values for cations and positive values for anions and fixed y-coordinates from a file to get a 6-vertice-polygon with cations Mg, Ca, Na+K in meq/L on the left side and the anions SO_4, HCO_3, Cl in meq/L on the right side. But there is already Open Source Software as Python scripts from [Waterloo 2014] that can not only plot series of Stiff-diagrams but also (rectangular) Piper-diagrams. A Python environment with many scientific libraries (NumPy, SciPy,...) included like "Python(x,y)-2.7.10.0a.exe" to run Python scripts is best but also ESRI ArcGIS has Python integrated and the (IDLE-)shell can also be used outside of ArcGIS. [Waterloo 2014] offers 2 Python scripts (stiff_fill.py and piper_rectangular.py) and a data file (watersamples.txt) that is read by the scripts. To use the scripts our data must be of 'watersamples.txt' structure and the script code must be adapted:

#Type	Cl	HCO3	SO4	NaK	Ca	Mg	EC	NO3	Sicc
1	1.72	4.02	0.58	1.40	4.53	0.79	672.00	0.40	0.21
2	0.90	1.28	0.54	0.90	1.44	0.74	308.00	0.36	0.56
2	4.09	4.29	0.38	3.38	4.74	0.72	884.00	0.08	0.15
4	4.26	4.34	0.25	3.44	5.22	0.38	904.00	0.19	0.29
2	1.21	2.02	0.64	2.16	1.88	0.00	403.00	0.16	-0.7
2	1.07	1.32	0.69	1.52	1.34	0.38	324.00	0.16	0.45

Fig. 29: 'watersamples.txt' [Waterloo 2014] w meq/L for Stiff- & Piper- diagrams with Python.

The first column (Type) determines the color of a Stiff diagram. The data needed is mainly in IC-results. Si^{4+} is taken from photometer results. Values for Na+ and K+ must be added. All ion concentrations (in mg/L) must be divided by their molar weight (in mg/mmol = g/mol) to get molar concentrations (in mmol/L and for univalent ions also meq/L). For SO_4^-, Ca^{++} and Mg^{++} we must multiply this with valence=2, for Si with valence=4. HCO_3 is missing but can be calculated from alkalinity by multiplying with 0.02. Every mol $CaCO_3$ can produce 2 mol HCO_3:

$$CaCO_3 + H_2O + CO_2 \leftrightarrow Ca^{++} + 2\,HCO_3^- \quad [6] \Rightarrow$$
1 mg/L alkalinity = 1mg mmol/(100mg L) $CaCO_3$ = 0.01 mmol/L $CaCO_3$ = 0.02 meq/L HCO_3^-

From the IC-results, alkalinity, and EC we get the following spreadsheet for Darmbach. The first column' number selects the diagram color (1-5):

meq/L = mg/L * valence/MolarWeight

Cl: * 1 /35.5 mg/mmol
HCO3:* 1 / 61 mg/mmol
SO4: * 2 / 96 mg/mmol
Na: * 1 / 23 mg/mmol
K: * 1 / 39 mg/mmol
Ca: * 2 / 40 mg/mmol
Mg: * 2 /24.3 mg/mmol
NO3: * 1 / 62 mg/mmol
Si: * 4 / 28 mg/mmol

	A	B	C	D	E	F	G	H	I	J
1	#Type	Cl	HCO3	SO4	NaK	Ca	Mg	EC	NO3	Sicc
2	1	0.28028	3.9	0.46317	0.5012	2.9492	0.45926	383	0.003532	0.032857
3	2	0.28332	3.84	0.50471	0.49887	3.6206	0.52082	449	0	0.994286
4	3	0.16904	3.7	0.51758	0.43346	3.28205	0.50971	410	0	0.975714
5	4	0.1238	3.6	0.39931	0.40098	2.981	0.47564	405	0	0.975714
6	5	0.12817	2.84	0.40156	0.42542	2.69685	0.39086	340	0.002048	0.45
7	2	0.12963	4.2	0.54146	0.29652	4.02795	0.60889	482	0.151371	1.08

Fig. 30: Darmbach meq/L of ions relevant for Python scripts.

I exported the spreadsheet as tabulator-separated text file 'waterlevel.txt' and adopt the Python program 'stiff_fill.py', because we have only 6 (instead of 18) locations, want only 2 instead of 5 diagrams per row and we comment out (with a #-sign) the figlegend() function:

```
'''
#nosamples = 18 # # number samples
nosamples = 6
ncol = 2 # number of columns of subplot
nrow = 3 # number of rows of subplot
...
# Legend for Figure
#figlegend((h1,h2,h3,h4,h5,h6,h7),('Dinkel',...
```

The program draws 6 Stiff diagrams in 3 rows of 2 columns each. 1St diagram = upper left corner = Post-Woog. Last diagram = lower right corner = Spring. The squares represent Si and the circles represent NO3. Zero meq/L is a white vertical line:

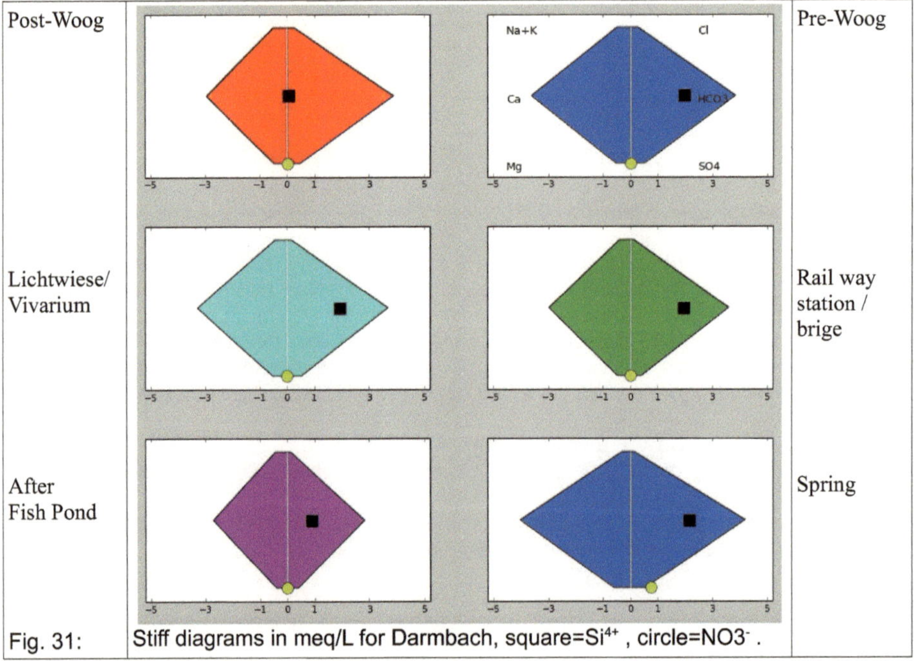

Fig. 31: Stiff diagrams in meq/L for Darmbach, square=Si^{4+} , circle=NO3$^-$.

5.3 Piper diagram

Piper diagrams can be produced with the GW_Chart program from USGS (that is otherwise used to calculate water budgets) and with a Python script (needs a Python interpreter and some Python libraries that add Matlab- and GNUplot- functionality to Python – all included e.g. in "pythonxy") from [Waterloo 2014]. Nearly all other water science diagram programs seem to be commercial (i.e. expensive and cumbersome in license handling).

GW_Chart asks for CO_3 and HCO_3 . The later that must be calculated from alkalinity with equation [6]. GW_Chart can work with meq/L, mg/L, %. For CO_3 a calculator opens within GW_Chart and asks for temperature, $[HCO_3]$ and pH to calculate $[CO_3]$. The problem is that we start with an alkalinity to calculate bicarbonate. GW_Chart calculates carbonate and both together are equivalent to a bigger alkalinity than that we started with. That means we will need some nasty iterations until both sum up again to the alkalinity we started with. Another problem is, that GW_Chart also needs the amount of total dissolved solids – TDS. Here is an example with mg/L (so GW_Chart must convert itself to meq/L) and one line for 1 location, but we can also write many lines for many locations to get them all in one diagram:

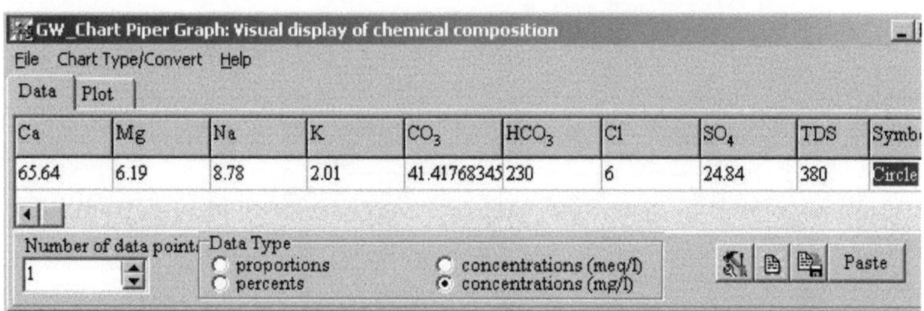

Fig. 32: Input for Piper diagram in GW_chart.

Instead of calculating all Darmbach values and inserting them into GW_Chart I try the Python alternative 'piper_rectangular.py' within the full-fledged "pythonxy" package, that has many libraries installed that are not all available in Python of ESRI ArcGIS. Installing one Python version complaines about others already installed. So I tried "pythonxy" on a different PC without ArcGIS. The test data has only 5 different water types (at 18 locations) and we have 6 locations that should be treated as 6 different water types. Therefore we must add in some places of the programm where type 1 to 5 are treated the 6th by just copying the 5th and replacing the 5 by a 6. Additionally the legend of the diagram (figlegend()) is adapted to our 6 stations along Darmbach. This creates the Python program 'piper_rectangular2.py':

```
...
  if obs[i, 0] == 5:
      ax1.plot(100*obs[i,Ca]/(obs[i,EC]/100), 100*obs[i,Mg]/(obs[i,EC]/100), 'm^', ms = markersize)
  if obs[i, 0] == 6:
      ax1.plot(100*obs[i,Ca]/(obs[i,EC]/100), 100*obs[i,Mg]/(obs[i,EC]/100), 'bs', ms = markersize)
...
  if obs[i, 0] == 5:
      h5,=plot(100*obs[i,Cl]/(obs[i,EC]/100), 100*obs[i,SO4]/(obs[i,EC]/100), 'm^', ms = markersize)
  if obs[i, 0] == 6:
      h6,=plot(100*obs[i,Cl]/(obs[i,EC]/100), 100*obs[i,SO4]/(obs[i,EC]/100), 'bs', ms = markersize)

  if obs[i, 0] == 5:
      h5,=ax2.plot(100*obs[i,NaK]/(obs[i,EC]/100), 100*(obs[i,SO4]+obs[i,Cl])/(obs[i,EC]/100), 'm^', ms = markersize)
  if obs[i, 0] == 6:
      h6,=ax2.plot(100*obs[i,NaK]/(obs[i,EC]/100), 100*(obs[i,SO4]+obs[i,Cl])/(obs[i,EC]/100), 'bs', ms = markersize)

figlegend((h1,h2,h3,h4,h5,h6),('Post-Woog','Pre-Woog','Vivarium','Rail way','Fish Pond','Spring'), \
...
```

Markers are rs=red square for Post-Woog, b>=blue arrow to the right for Pre-Woog, c<=cyan arrow to the left for Lichtwiese/Vivarium, go= green circle for the rail way station, m^=magenta arrow up for the Fish Pond and bs=blue square for the spring. I open the the Python program in the SciTE-editor of "pythonxy" where it is started with "Go" in the Tools-menu. It plots the following graph that is also stored in 3 different graphic formats (eps, png, gif):

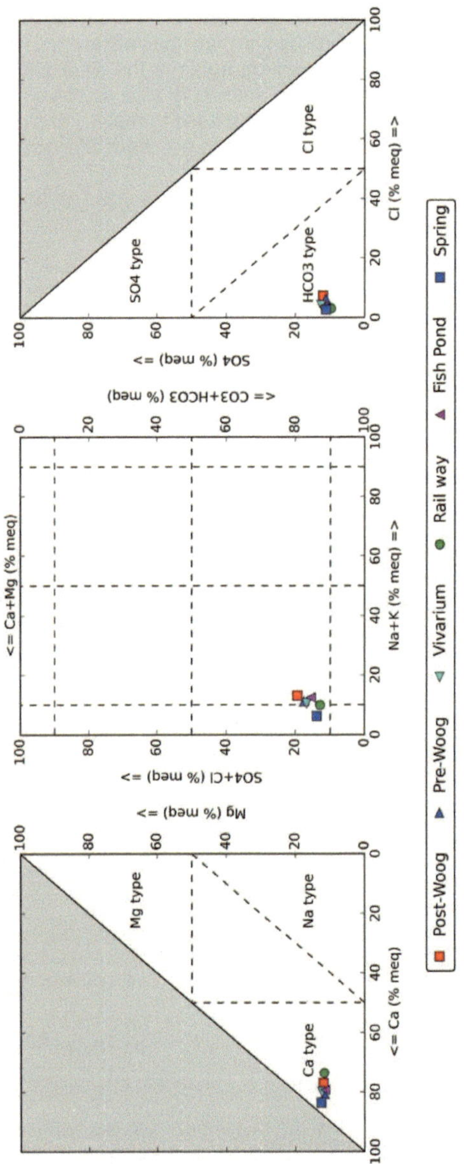

Fig. 33: Piper diagram for Darmbach from Python script.

5.4 Discussion of Darmbach water properties

As we see from the last (Piper) diagram, all our water is of type Ca / Carbonate – it contains a lot of calcium and (bi)carbonate ions.

In lakes Ca drops: Spring/After-Fish-Pond = 81/54 and Pre-Woog/Post-Woog = 72/59. Also Si drops in lakes: Spring/After-Fish-Pond = 7.6/3.15 and Pre-Woog/Post-Woog = 6.96/0.23. Ca and Si may just precipitate in the lakes due to low flow speed or also used up by living creatures for building shells. Si concentration is high due to the crystalline rocks of Odenwald mountains.

Concentrations of Na (6 .. 10) and Cl (4.6 .. 10) increase from Spring to Post-Woog probable by intake of salt (NaCl) used e.g. on roads and footpaths to conquer snow and ice in winters traveling slowly from road ditches to streams like Darmbach.

NO2 increases from Spring (0.003) to Pre-Woog (0.071) by exhaust emissions from burning hydro carbons but is decreased strongly in lakes (Woog and Fish Pond).

NO3 is contained in soil and in higher concentrations in fertilizers. The highest NO3 concentration is in the spring water but it is much lower that the allowed limits.

Br seems quite high but IC machine has a carbonate peak close to Br, so the Br values may be wrongly reported by the machine.

6 References

- Domenico, P.A., Schwartz, F.W., 1990: "Physical and Chemical Hydrogeology", John Wiley & Sons, 824 pp.
- Hem, J.D., 1985: "Study and Interpretation of the Chemical Characteristics of Natural Water", USGS, paper 2254, 272 pp.
- Janert, P.K., 2010: "Gnuplot in action – understanding data with graphs", Manning Publications, Greenwich, 360 pp.
- Marandi, A., 2014: "Lecture Hydrogeology I – 3406 - for MSc TropHEE", Technische Universität Darmstadt
- Waterloo, M.J., Post, V.,E.A., 2014: "Python programming for Hydrology MSc students", VU (Vrije=Free) University Amsterdam, 64 pp. http://python.hydrology-amsterdam.nl/ and https://code.google.com/p/pythonxy/wiki/Downloads
- Schiedek, T., 2014: "Lecture Water Analysis – 3214 – for MSc TropHEE", Technische Universität Darmstadt
- Schwedt, G., 2007: "Taschenatlas der Analytik", John Wiley-VCH, 249 S.
- Uliana, M.M., 2012: "Hydrogeology Lecture Notes, Ed 2.3", 139 pp.
- Winter, T.C., Harvey, J.W., Franke, O.L., Alley, W.M., 1998: "Ground Water and Surface Water – A single resource", USGS, Circular 1139, 87 pp.